Exploring Probability

Exploring Probability was prepared under the auspices of the American Statistical Association—National Council of Teachers of Mathematics Joint Committee on the Curriculum in Statistics and Probability.

This book is part of the Quantitative Literacy Project, which was funded in part by the National Science Foundation.

Exploring Probability

Claire M. Newman
Queens College

Thomas E. Obremski
University of Denver

Richard L. Scheaffer
University of Florida

DALE SEYMOUR PUBLICATIONS

Cover Design: John Edeen and Francesca Angelesco
Technical Art: Colleen Donovan
Illustrations: Deborah Morse
Editing and Production: Larry Olsen

Printed in the United States of America. Published simultaneously in Canada.

This publication was prepared as part of the American Statistical Association Project—Quantitative Literacy—with partial support of the National Science Foundation Grant No. DPE-8317656. Any opinions, findings, conclusions, or recommendations expressed in this publication are those of the authors and do not necessarily represent the views of the National Science Foundation. These materials shall be subject to a royalty-free, irrevocable, worldwide, non-exclusive license in the United States Government to reproduce, perform, translate, and otherwise use and to authorize others to use such materials for Government purposes.

ISBN 0-86651-333-7
Order Number DS01701

fghijkl—MA—9 5 4 3 2 1 0

DALE
SEYMOUR
PUBLICATIONS
P.O. BOX 10888
PALO ALTO, CA 94303

CONTENTS

PREFACE

This is the second in a series of publications produced by the ASA-NCTM Joint Committee on the Curriculum in Statistics and Probability. The series includes *Exploring Data, The Art and Techniques of Simulation,* and *Exploring Surveys and Information from Samples.* These four units cover the basic concepts of statistics and probability. The approach emphasizes use of real data, active experiments, and student participation. There are no complicated formulas or abstract mathematical concepts to confuse or distract you.

Exploring Probability covers elementary probability by using only counting skills and some knowledge of fractions. Complicated counting algorithms such as combinations and permutations are not introduced here. The unit provides the background for the many practical applications of probability discussed in *The Art and Techniques of Simulation.*

The material is designed to give you a working knowledge of basic probability.

I. INTRODUCTION

In our daily conversations, it is common to speak of events in terms of their chances of occurring. We speak of the chance that our team will win the big game, the chance that we will get an A in mathematics, or the chance of being elected to a school office. The word *probability* is sometimes used in place of the word *chance*. Then we might speak of the probability of getting a hit in a baseball game or the probability that our new puppy will be born male.

The terms *chance* and *probability* are usually applied to those situations for which we cannot completely determine the outcome in advance. We are doubtful of what will happen. We do not know if our team will win or lose, so we talk about the team's chance of winning. We are not sure if we will get an A in mathematics, so we talk about our chance of getting an A. In other words, we are *uncertain* about what will actually happen when we use the term *chance*. There are, however, situations for which we know the outcomes in advance. We are sure that the sun will rise tomorrow and that school will begin again next fall. We say that we are *certain* about such events. All of us have often heard such statements as "I am certain that I will pass this year" or "I am certain that I can find my way home."

Sometimes we are not certain about how a situation will turn out, but we think the chances are good that a particular outcome will occur. We may call such an outcome *likely* or *highly likely*. If I am confident about knowing the answers to the questions on a math test, I could say that it is likely that I will pass the test. If you are fairly sure that you will go to a movie next Saturday, then you could say that you are likely to attend a movie. However, if we think the chances that an event will occur are low, then we may call such an event *unlikely*. If your team is missing two of its star players, then it may be unlikely that it will win. So, *likely* refers to those events that have high probability of occurring, and *unlikely* refers to those events that have low probability of occurring.

The concept of chance, or probability, is widely used in the natural and social sciences because few results in these areas are known in advance absolutely. Most events are reported in terms of chances—for example, the chance of rain tomorrow, the chance that you can get home from school or work safely, the chance that you will live past the age of 60, the chance of getting a certain disease (or recovering from it), the chance of inheriting a certain trait, and the chance of a candidate winning an election.

Of what use are these probabilities? Perhaps the most important use is to help us make decisions as we go through life. If a student knows that his or her chance of getting an A in mathematics is low, then he or she may decide to study harder. If a certain medicine has only a low chance of curing an illness you may have, then you will probably not waste your money on that particular medicine. If rain is likely, you will be inclined to carry an umbrella or take your raincoat, but if rain is unlikely you will probably not bother with these extra articles. Businesses and industries make important decisions using similar reasoning. For example, insurance companies are interested in the probability of auto accidents among persons in certain age groups, and industries are interested in the probability that a new product will make a profit or the probability that an item can be manufactured without defects in workmanship.

To introduce you to some situations involving probability, think about each of the following events. How likely is each to occur? Use the accompanying scale to assign a number from 0 to 1 to each event, with 0 representing *impossibility* and 1 representing *certainty*.

1. You will be absent from school at least one day during this school year.

2. You will have cereal for breakfast one day this week.

3. You will have cereal for lunch one day this week.

4. It will snow in your town in July.

5. It will rain sometime in July in your town.

6. The sun will rise tomorrow.

7. A person can live without water for two months.

8. A Democrat will win the next presidential election.

9. The next baby born in your local hospital will be a boy.

10. You will get an A in your next math test.

Now that you have given some thought to probable and improbable events, we shall explore a more systematic approach to probability in the next section.

Application 1

The Spinner

1. A spinner is divided into areas labeled *red* and *white*, as in the accompanying diagram.

   a. If you were to spin the spinner, would you be just as likely to obtain red as white? If not, which color is more likely to occur? Why?

   b. Are you certain of getting at least one red in 100 spins?

   c. Is it very likely that you will not spin any reds in 100 spins?

   d. Is it possible never to spin a red in 100 spins?

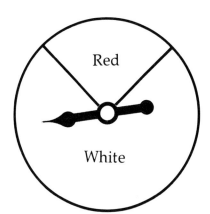

2. Think about events that may occur in your life.

   a. List three events that are certain.

   b. List three events that are impossible—that is, they cannot occur.

   c. List three events that are highly likely.

   d. List three events that are unlikely.

II. EXPERIMENTING WITH CHANCE

We have seen that, in our daily experiences, there are many situations that involve chance outcomes. These are situations for which we do not know the outcome in advance, but we may be able to list possible outcomes. We will now investigate some simple situations of this type by actually observing how often specific outcomes occur. Our goal is to assign *numbers* to various events of interest. These numbers will represent the *chances*, or *probabilities*, of the events we study.

Think about the following simple situation. An official is to toss a coin at the start of a football game to determine who gets first chance to receive the ball. Will the coin come up heads *(H)* or tails *(T)*? We cannot know the outcome in advance, but we do know that there are only two possible outcomes. We can study the possible outcomes of this coin-tossing experiment by actually tossing a coin repeatedly and observing what happens. Throughout our discussion of probability, the term *experiment* will refer to any activity we use to produce observable data. Any *set* of the possible outcomes of an experiment is called an *event*. Let us conduct an experiment to see how coin-tossing works.

Application 2

Tossing a Coin

Toss a coin 50 times. Record the results in the accompanying table.

	Tally	Total
Heads		
Tails		
Total		50

1. Calculate the fraction of tosses that were heads. Use the formula

$$\text{fraction of heads} = \frac{\text{total number of heads}}{\text{total number of tosses}}$$

2. Calculate the fraction of tosses that were tails.

3. Compare your results with those of other students in the class. Do you find some variation in the results?

4. Are the results of this experiment what you would have expected to observe?

The fraction of heads obtained in the coin-tossing experiment is called the relative frequency. The relative frequency can now be used to *estimate* the *chance*, or *probability*, of observing a head in the future. We will say that this fraction is an *estimate* of the probability of obtaining a head, based on the 50 tosses we observed. We may have counted 16 heads in 50 tosses. Our estimate of the probability of heads is then $\frac{16}{50}$. You may have obtained a different number of heads. Your estimate of the probability of observing a head is then different from $\frac{16}{50}$. Upon checking with other students in the class, you may find a large number of different estimates of the probability of observing a head.

Estimates of probabilities usually change from experiment to experiment. However, if each student in the class had tossed the coin a *large* number of times (say, 1,000), then the estimates of the probability of heads would not vary quite so much. All estimates, in that case, should be quite close to one another, and any one of them would provide a good estimate of the probability of observing a head.

You have, no doubt, already suspected that the fraction of heads observed in coin-tossing should be close to $\frac{1}{2}$. There are, however, many situations in which an estimate of probability is nearly impossible without some experimentation. Suppose you want to know the probability that a car approaching a specific intersection near your home will turn left. You must observe a number of cars and count the number of left turns. These observations of cars make up the experiment in this case. If, out of 50 cars observed, 20 turn left, then $\frac{20}{50}$ = 0.40 (or 40 percent) is the relative frequency or fraction of left turns. The relative frequency is used as the estimate of the probability that a car will make a left turn. If 200 cars go through this intersection tomorrow, you would *expect* 40 percent of them (80 cars) to turn left.

For some problems, the experiment has already been conducted, and we have only to look at the data. Suppose Joe is coming up to bat in a baseball game. What is the chance he will get a hit? Here we need to observe the fraction of times he has gotten a hit in the past, but this information is already collected and summarized in his batting average. If Joe has been to bat 10 times and has 3 hits, then his batting average is $\frac{3}{10}$ = 0.300. We say he is "hitting 300." This is really a probability measure. Joe's chance of getting a hit his next time at bat is 0.3. In other words, we expect Joe to get a hit on 30 percent of his official turns at bat. If Joe comes to bat 10 times in the next two games, we expect him to get 3 hits. Of course, he may get no hits or many more than 3 hits, but we expect him to average about 3 hits for every 10 times at bat.

It should now be clear that the probability of an event, *E*, can be estimated by

$$P(E) = \frac{\text{the number of observations favorable to } E}{\text{the number of observations in the experiment}}$$

To review, if 50 tosses of a coin result in 16 observations of heads, then the estimate of the probability of observing a head on a future toss is $\frac{16}{50}$ = 0.32, or 32 percent.

Now try the following Applications, which are designed to help you become more familiar with probability experiments and how they are used.

Application 3

The Spinner

Construct a spinner like the one shown below, with eight sectors of equal size. (A larger version of this spinner base is provided at the end of the book. A pointer can be made by spinning a paper clip around a pencil point.) Number each sector and color it yellow, red, or blue as shown. Spin the spinner 30 times and record the *color* on which the spinner lands each time. If the spinner lands on a line, spin again.

	Red	Blue	Yellow
Number Observed			

Use your data to answer the following questions.

1. Would you say that yellow is more or less likely than red? Why?

2. Would you say that blue is more or less likely than red? Why?

3. Estimate the probability of the spinner landing on yellow the next time it is spun.

4. Estimate the probability of the spinner landing on blue the next time it is spun.

5. If you spin this spinner 90 more times, about how many times would you expect it to land on blue?

6. Combine your numbers with those of the rest of the class, and write down the total number of reds, blues, and yellows seen by the entire class. Now, answer questions 1 through 5 again.

Keep your spinner and these results for further experiments.

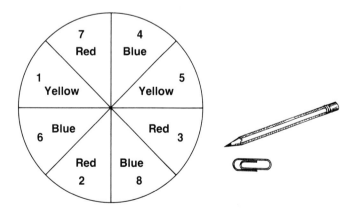

Thumbtack Toss

Place 20 identical thumbtacks in a paper cup. Shake the cup and roll (gently) the tacks onto the desk. Count the number of tacks pointing up (the point does not touch the desk) and the number pointing down (the point touches the desk). Record the data in the boxes below. (*Note:* To keep the tacks under control, place them in a clear plastic cup and cover the top with clear plastic held by a rubber band. Shake the cup and turn it upside down on a desk. You can count the number of tacks pointing up or down by looking through the clear plastic.)

Number of tacks pointing up	
Number of tacks pointing down	
Total	20

Use your data to answer the following questions.

1. Is "point up" more likely or less likely than "point down"?

2. If one tack is to be tossed onto the desk, estimate the probability that it will point up.

3. If a cup containing 100 tacks has its contents rolled onto the desk, how many tacks would you expect to point down?

4. Combine your results with those of four other students so that you have data on 100 tacks. Answer questions 1, 2, and 3 again. Are there any differences from your earlier answers?

Application 5

Tossing a Die

Work with a partner as a team. Toss a standard six-faced die (*die* is the singular of *dice*) 60 times and observe the number of dots on the top face. One person should toss the die, and one should record the results in the accompanying table.

	No. of ⚀	No. of ⚁	No. of ⚂	No. of ⚃	No. of ⚄	No. of ⚅
Tally						
Total						

Use your data to answer the following questions.

1. Did you get all six numbers? Would you expect that, for 60 tosses, each number would come up at least once? Explain.

2. Estimate the probability of tossing a 5 with your die.

3. Estimate the probability of tossing a 4 with your die.

4. Are the answers to questions 2 and 3 nearly equal?

5. Estimate the probability of tossing an even number with your die. How does this answer compare with the answer to question 3?

6. Estimate the probability of tossing a number larger than 4.

7. If you tossed your die 100 more times, about how many 4's would you expect to see among the 100 tosses?

Now, ten teams should enter their results on the Combined Data Table, on page 10.

	Combined Data Table						
Team	No. of [⚀]	No. of [⚁]	No. of [⚂]	No. of [⚃]	No. of [⚄]	No. of [⚅]	Total No. of Tosses
A							
B							
C							
D							
E							
F							
G							
H							
I							
J							
Total							600

8. Estimate the probability of tossing a 5, using the combined data. How does your answer compare with the answer for question 2?

9. Estimate the probability of tossing a 4, using the combined data. How does this answer compare with the answer for question 3?

10. Estimate the probability of tossing an even number, using the combined data. How does this answer compare with the answer for question 5?

Guessing Numbers

Get your pencil ready to write down a number. Ready? Without hesitating, write down a number from 1 to 4.

Fill in the boxes below with the number of students in your class who responded 1, 2, 3, and 4.

Number of Students Choosing Each Number

1	2	3	4

1. What is the total number of responses for your class?

2. Which number was written down most frequently?

3. Studies have shown that, from among the numbers 1, 2, 3, 4, the number 3 is chosen most frequently. If we want to estimate the probability that a person not in your class would choose number 3, how would we use the information in the boxes to do so?

4. Now estimate the probability of students writing down the responses 1, 2, and 4.

Counting Students

For this study, we will provide the data. You may want to conduct a similar study by interviewing the students in your school. Suppose a group of 20 students randomly selected from your school contains 12 girls and 8 boys. (You can select 20 students at random by placing the names of all students in a box, mixing them, and drawing 20 names.) Four of the girls and three of the boys live within easy walking distance of the school. The data may be displayed as follows:

	Girls	Boys	Total
Lives within walking distance	4	3	7
Does not live within walking distance	8	5	13
Total	12	8	20

What does the number 5 in the table mean? It means that, of the 20 students, there are 5 boys who do *not* live within walking distance of the school.

Another student, not among those interviewed, is assigned a locker next to yours.

1. Estimate the probability that this student is a girl.

2. Estimate the probability that this student lives within walking distance of school.

3. A reasonable answer to question 1 is $\frac{12}{20}$ or $\frac{3}{5}$, since there are 12 girls and 20 students in the sample. Now find the probability that the person named is a girl who lives within walking distance of school. (*Hint:* What does the number 4 represent in the table?)

4. Suppose there are 1,000 students in the school. How many would you expect to live within walking distance of the school?

5. If the person assigned a locker next to yours is known to be a girl, estimate the probability that she lives within walking distance of school.

Application 8

Generating Your Own Data

Select the names of two popular movies. Call one movie A, and the other, B. Now interview a number of students in your school and find out how many have seen each movie. (The interviewed students should be randomly selected from all students in the school, as described in Application 7.) Record the numbers in the following table:

	Seen B	Not Seen B	Total
Seen A			
Not Seen A			
Total			

Now select another student in your school whom you have not yet interviewed. Based on the data you have for the students you have interviewed,

1. Estimate the probability he or she has seen A.

2. Estimate the probability he or she has seen B.

3. Estimate the probability he or she has seen neither.

4. Among a group of 100 students from your school, none of whom were interviewed by you, how many students would you expect to have seen both movies?

Causes of Fires

The data for this experiment are provided below. These are real data, and you may construct a similar experiment by looking for graphs and other data displays in a newspaper or magazine.

Number of Fires of Various Types Among 100 Typical Home Fires	
Cause of Fire	Number Reported
Heating system	22
Cooking	15
Electrical system	8
Smoking	7
Appliances	7
Other	41

Source: National Fire Incident Reporting Service, 1978.

Suppose a fire starts in a home down the street from where you live. Using the table above, estimate the probability that:

1. It is a cooking fire.

2. It started in the electrical wiring.

3. It was *not* caused by smoking.

4. It was caused by the heating system or appliances.

5. It was caused by someone cooking or smoking.

6. Can we estimate the probability that the fire was caused by lightning?

Application 10

Accident Statistics

The following table lists the total number of accidental deaths in the United States in 1979 and the number of fatal motor-vehicle accidents.

Accidental Deaths in the United States, 1979		
	All Types	Motor Vehicle
All ages	105,312	53,524
Under 5	4,429	1,461
5 to 14	5,689	2,952
15 to 24	26,574	19,369
25 to 44	26,097	15,658
45 to 64	18,346	8,162
65 to 74	9,013	3,171
75 and over	15,164	2,751
Male	74,403	39,309
Female	30,909	14,215

Source: *The World Almanac & Book of Facts,* 1979 and 1984 edition, copyright © Newspaper Enterprise Association, Inc. 1978 and 1983, New York, NY 10166.

You are told that a certain person recently died in an accident. Estimate the probability that:

1. It was a motor-vehicle accident.

2. The person was male.

3. It was a motor-vehicle accident, assuming that the person was a male.

4. It was a motor-vehicle accident, assuming that the person was between 15 and 24 years of age.

5. It was a female who was involved in an accident that was not a motor-vehicle accident.

Application 11

Smoking Data

The accompanying graphs give information about smoking among males and females in 1965 and 1980.

1. Suppose you were to meet Ms. J by chance for the first time. What is your estimate of the probability that:

   a. she does not smoke?

   b. she has never smoked?

   c. she has smoked sometime in her life?

2. Suppose a man had been randomly selected during the year 1965 as part of an experimental study. What is your estimate of the probability that he was a smoker? Is it more or less likely that such a selection would produce a smoker today?

3. Construct two probability questions of your own that can be answered using the graphs.

Quitting Proves Hard

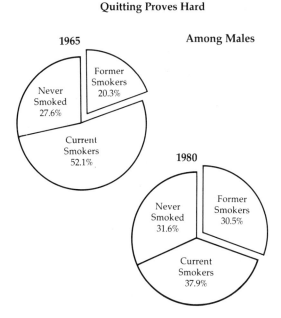

Among Males

1965
- Never Smoked 27.6%
- Former Smokers 20.3%
- Current Smokers 52.1%

1980
- Never Smoked 31.6%
- Former Smokers 30.5%
- Current Smokers 37.9%

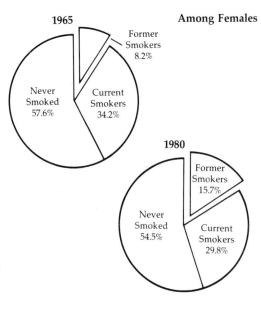

Among Females

1965
- Never Smoked 57.6%
- Former Smokers 8.2%
- Current Smokers 34.2%

1980
- Never Smoked 54.5%
- Former Smokers 15.7%
- Current Smokers 29.8%

Source: Public Health Service; *The New York Times*, Dec. 25, 1984. Copyright © 1984 by The New York Times Company. Reprinted by permission.

Application 12

Marital Status of the Unemployed

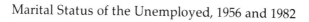

Marital Status of the Unemployed, 1956 and 1982

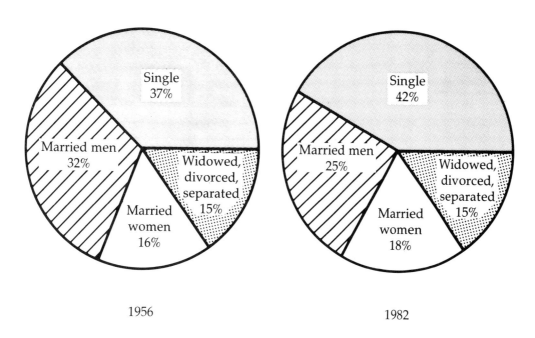

1956   1982

Source: "Workers Without Jobs: A Chartbook on Unemployment,"
Bureau of Labor Statistics, 1983.

The above graphs show the marital status of the unemployed for 1956 and 1982.

1. If you had met an unemployed worker in 1956, what was the probability that this person would have been a married woman? What is the probability that an unemployed worker in 1982 was a married woman? Can you suggest reasons for the change?

2. Suppose your city had 1,000 unemployed workers in 1982. How many would you expect to have been:

   a. married men?

   b. single?

   c. married women?

The Adult Population of the United States

The following data from the U.S. Census Bureau show the percentages of the adult population under age 65 by age classes. The data are for 1981.

Adult Population of the U.S. by Age Classes	
Age Class	Percentage of Adult Population Under Age 65
18-24	21
25-34	28
35-44	19
45-64	32

Use the data to answer the following questions.

1. If you meet a stranger known to be between 18 and 64 years old, what is your estimate of the probability that the stranger is:

    a. between 18 and 24 years old?

    b. over 34 years old?

    c. under 35 years old?

2. If a large firm employs 10,000 workers between the ages of 18 and 64, how many of them would you expect to be:

    a. between 18 and 24 years old?

    b. over 34 years old?

III. KNOWING OUR CHANCES IN ADVANCE

In the coin-tossing experiment in Application 2, you no doubt expected to obtain heads on about $\frac{1}{2}$ of the tosses. This seems reasonable, since there are only two possible outcomes to each toss of the coin, and these outcomes seem to be equally likely. In fact, we say that the *probability* of a head occurring on a toss of a coin is $\frac{1}{2}$, and we write this expression as

$$P(H) = \frac{1}{2}$$

Of course, the probability of a tail occurring is also $\frac{1}{2}$.

In the case of tossing a die, there are six equally likely outcomes. Thus, the probability of observing a 3 on a single toss of a die is $\frac{1}{6}$. Suppose we are interested in the event "observe an even number." Three of the six possible outcomes are favorable to this event, namely, the outcomes 2, 4, and 6. Therefore, the probability of observing an even number is $\frac{3}{6}$ or $\frac{1}{2}$.

Whenever we have an experiment with equally likely outcomes, we may define the probability of an event, E, as

$$P(E) = \frac{\text{number of outcomes favorable to } E}{\text{total number of possible outcomes}} = \frac{f}{n}$$

We use this definition, $P(E) = \frac{f}{n}$, only when we are dealing with *equally likely outcomes*. It will not work with outcomes that are not equally likely. For example, the thumbtack toss of Application 4 shows that the two possible outcomes "tack points up" and "tack points down" are not equally likely. Therefore, the probability of the tack pointing down is *not* $\frac{1}{2}$.

The probabilities determined by the rule given in this section are called *theoretical* probabilities because they are determined by what should happen, ideally, in situations with equally likely outcomes. The key to finding these theoretical probabilities is to list *all* possible equally likely outcomes. Then the theoretical probability of any event can be found by simply counting the number *(f)* of outcomes favorable to an event E, counting the total number *(n)* of equally likely outcomes, and using the formula $P(E) = \frac{f}{n}$.

The *theoretical probability* of observing a head in a toss of a coin is $\frac{1}{2}$. But when we actually toss a coin (see Application 2) 50 times, we usually do not get exactly 25 heads. The relative frequency of heads for the 50 tosses of a coin (obtained experimentally) will vary from student to student—and the relative frequency is the *estimated probability* of observing a head. If, however, the number of tosses were 1,000 instead of 50, the relative frequency should be quite close to $\frac{1}{2}$ and should vary little from student to student. The *theoretical probability* of $\frac{1}{2}$ can be used to predict what will happen in the *long run*. The theoretical probability is roughly the same as the estimated probability obtained from a large number of tosses of the coin. If we can determine the theoretical probability of an event, we generally use it as the measure of that event's probability. When we cannot determine the theoretical probability, we estimate the probability by conducting an experiment and determining the relative frequency of the desired event.

The following Applications allow you to use the definition of theoretical probability and to explore the relationship between theoretical and estimated probability.

Application 14

The Spinner Revisited

Obtain the spinner you used in Application 3. The eight numbered sections should be of approximately equal size. Thus, the eight possible numbered outcomes should be equally likely. Using our definition of theoretical probability:

1. Find *P(spinner landing on 3)*.

2. Find *P(spinner landing on an even number)*.

3. Find *P(spinner landing on blue)*.

Spin the spinner 40 times and record the frequency of outcomes in the boxes below.

	1	2	3	4	5	6	7	8	Total
Number of Times Observed									40

4. Find the fraction of 3's observed in the total sample of 40 spins. Compare this relative frequency with your answer to question 1.

5. Find the fraction of even numbers observed in the sample. Compare the result with your answer to question 2.

6. Find the relative frequency of blues observed in the sample. Compare this fraction with your answer to question 3.

7. Questions 4, 5, and 6 give estimates of probabilities. How do you think such estimates would compare with the theoretical probabilities in questions 1, 2, and 3 if we had data on 4,000 spins instead of 40 spins?

Application 15

Play Your Cards Right

Obtain a standard deck of 52 playing cards. Mix them well and count out 25 cards WITHOUT LOOKING AT THEM. Put aside the remaining cards. You are going to perform an experiment to estimate the probability of drawing a club, a diamond, a heart, and a spade from your deck of 25 cards.

A. Mix the 25 cards well. Draw one card. Record its occurrence in the appropriate box below.

B. Replace the card and shuffle the deck of 25 cards.

C. Draw another card and record its suit.

D. Repeat the above steps until you have recorded a total of 25 draws.

Clubs	Diamonds	Hearts	Spades	Total
				25

Use your data to answer the following questions.

1. What is your estimate of the probability of drawing a club from the deck of 25 cards?

2. What is your estimate of the probability of drawing a diamond from the deck of 25 cards?

3. What is your estimate of the probability of drawing a heart from the deck of 25 cards?

4. What is your estimate of the probability of drawing a spade from the deck of 25 cards?

5. Now look at your 25 cards. Count the number of cards in each suit.

6. What is the theoretical probability of obtaining a club? a diamond? a heart? a spade?

7. How do these theoretical probabilities compare with the estimated probabilities obtained in the experiment?

8. Suppose you had recorded a total of 2,500 draws (instead of 25) in your experiment. How would the estimated probabilities compare with the theoretical probabilities then?

Birthdays

If it is not the month of January, take the January page from a large calendar and cut out all of the 31 numbers. Be sure the cut-out numbers are all close to the same size. Place them in a large container and shake them up. Suppose one number is picked out of the container. Before actually selecting a number, find the probability that:

1. The number is 17.

2. The number is an odd number.

3. The number is a one-digit number.

Now draw a number from the container and write it down. Return the number to the container, shake it up, and draw again. Repeat this process 20 times.

4. Find the fraction of odd numbers among the 20 numbers selected. Compare your answer with your answer to question 2.

5. Find the fraction of one-digit numbers among the 20 numbers selected. Compare your answer with your answer to question 3.

6. What is your birth month and birthday? For example, if your birthday is August 12, August is the month and 12 is the day of the month. What is the probability that one number selected from the container matches your birthday?

All the students in your class should have the same answer to the above question, namely, $\frac{1}{31}$. Why is this so?

7. Why is it important to the Application that all the cut-out numbers be about the same size?

Application 17

Experiment in ESP

Do you have ESP (extrasensory perception)? Try this experiment and see.

A. Make a set of 40 cards of the same size using four different symbols, so that you have ten cards for each symbol. They might look like this:

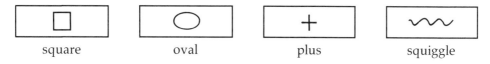

square   oval   plus   squiggle

B. Choose a partner. Ask him or her to face away from you (or blindfold him or her).

C. Mix the cards well.

D. Turn over a card and concentrate on the symbol it shows.

E. Ask your partner to read your mind and tell you what is written on the card.

F. Record the answer *without* telling him or her whether or not it is correct.

G. Repeat the procedure and tally the results until you have recorded a total of 20 answers.

Right Answer	Wrong Answer	Total

1. Do you think your partner has ESP? Why or why not?

2. If your partner is just guessing, what is the probability of his or her guessing correctly on any one trial?

3. If you were to run this experiment again for 100 trials, about how many answers do you predict would be correct?

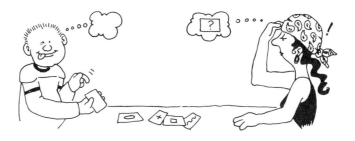

Random Numbers

1. Suppose the ten digits from 0 through 9 are written on ten poker chips, one number per chip. The chips are then placed in a box and mixed. You reach into the box and pull out one chip. What is the probability that the number on the chip is:

    a. a 1?

    b. even?

    c. greater than 7?

    d. divisible by 3? (Zero is divisible by 3.)

2. Suppose you pull out a chip that is *not* zero. (If the first chip you pull out is zero, replace the chip and repeat the process until you obtain a chip that is *not* zero.) Write down the number on your chip, put your chip back into the box, mix the chips, and then draw another one. Write the number on this second chip to the right of the first one, forming a two-digit number. What is the probability that the two-digit number:

    a. ends in a 3?

    b. is even? (Zero is an even number.)

Numbers selected in this way are called *random numbers*. Each number has the same probability of being selected. Random numbers are often used in selecting random samples. Suppose you have ten close friends. You could randomly select one of them to attend a ball game with you by numbering them from 0 through 9 and then choosing the one whose number is the same as the number you selected in question 1. We will use random numbers in future Applications.

Application 19

A Random Number Table

You should complete Application 18 before attempting this one. Suppose the experiment in Application 18 is repeated many times, and the resulting numbers are written in table form. The result would be a random number table. Such a table with 250 single-digit entries is shown below. The numbers are separated into five-column groups for ease of reading.

03222	39951	12738	50303	25017
87002	61789	96250	99337	14144
68840	94259	01961	42552	91843
88323	28828	64765	08244	53077
55170	71062	64159	79364	53088
84207	52123	88637	19369	58289
00027	43542	87030	14773	73087
33855	00824	48733	81297	80411
50897	91937	08871	91517	19668
21536	39451	95649	62556	23950

1. How many 1's would you expect to see in such a table?

2. How many even digits would you expect to see in such a table?

3. How many digits greater than 7 would you expect to see in such a table?

4. How many digits divisible by 3 would you expect to see in such a table?

5. Count the number of 1's, the number of even digits, the number of digits greater than 7, and the number of digits divisible by 3 in the table shown above. Are the frequencies close to what you expected?

IV. COMPLEMENTARY EVENTS AND ODDS

1. Complementary Events

Very often, when we consider an event and its probability, we are also interested in knowing the probability that the event will not occur. For each event, E, there exists the complement of that event, *not E* (which we will write as E'). We shall see that the probability of E and the probability of E' always have the same relationship to each other.

Study the probability of each event E below, and compare it with the probability of the event E' (that E will not occur). Each of these examples comes from an experiment discussed in the previous sections.

Experiment	E	$P(E)$	E'	$P(E')$
Toss of one coin	Head	$\frac{1}{2}$	Not a head	$\frac{1}{2}$
Spin of the spinner	Blue	$\frac{3}{8}$	Not blue	$\frac{5}{8}$
Toss of one die	Four	$\frac{1}{6}$	Not a four	$\frac{5}{6}$
Toss of one die	Greater than four	$\frac{2}{6}$	Less than or equal to four	$\frac{4}{6}$

Notice that $\frac{1}{2} + \frac{1}{2} = 1$, $\frac{3}{8} + \frac{5}{8} = 1$, $\frac{1}{6} + \frac{5}{6} = 1$, and $\frac{2}{6} + \frac{4}{6} = 1$. The sum of the two probabilities is always 1. That is, for each pair of complementary events,

$$P(E) + P(E') = 1$$

As another example, suppose the record of a weather station shows that its weather predictions have been accurate 89 times in the past 120 days. What is an estimate of the probability that its next forecast will be incorrect? Let C be the event "the forecast is correct," and let C' be the event "the forecast is incorrect." Then, $P(C) = \frac{89}{120}$. Since $P(C) + P(C') = 1$, $\frac{89}{120} + P(C') = 1$. It follows that

$$P(C') = 1 - \frac{89}{120} = \frac{(120 - 89)}{120} = \frac{31}{120}$$

In each case, we are dealing with a pair of fractions whose denominators are the same. The sum of the two numerators is always equal to that denominator.

$$\frac{1}{2} + \frac{1}{2} = 1 \qquad \text{and} \qquad 1 + 1 = 2$$
$$\frac{3}{8} + \frac{5}{8} = 1 \qquad \text{and} \qquad 3 + 5 = 8$$
$$\frac{1}{6} + \frac{5}{6} = 1 \qquad \text{and} \qquad 1 + 5 = 6$$
$$\frac{89}{120} + \frac{31}{120} = 1 \qquad \text{and} \qquad 89 + 31 = 120$$

In the spinner experiment, the probability of landing on yellow is $\frac{2}{8}$. What is the probability of not landing on yellow?

2. Odds

The numerators of each probability pair can be compared by means of a ratio. In the case of the toss of a die, we may say that the odds *in favor* of obtaining a 4 are 1:5, and the odds *against* obtaining a 4 are 5:1. Notice the order in which the probability numerators occur in each ratio. From $P(4) = \frac{1}{6}$ and $P(not\ 4) = \frac{5}{6}$, we obtain the odds *in favor* as 1:5 and the odds *against* as 5:1. In the spinner experiment, $P(blue) = \frac{3}{8}$ and $P(not\ blue) = \frac{5}{8}$. Therefore, the odds in favor of the spinner landing on blue are 3:5, whereas the odds against landing on blue are 5:3.

In the weather station example, what are the odds in favor of an accurate weather prediction? In the toss of a die, what are the odds against tossing a number greater than 4?

Observe that the odds in favor of obtaining a head in the toss of one coin are 1:1. In general, what does it mean to say that the odds in favor of an event are 1:1?

Application 20

Finding the Key

Suppose you have eight keys on your key ring. One opens the front door to your house, and one opens the back door. The other six will not help you get into your house. You come home after dark and randomly select a key.

1. What are the odds against the key opening your front door?

2. What are the odds against the key opening the back door?

3. What are the odds in favor of the key opening a door of your house?

4. You can rule out one key because it is much too small to be a house key, and you can tell that by touching it. Now answer questions 1, 2, and 3 again. Did the removal of one key change the odds considerably?

Application 21

Where the Students Go to School

Pupil Enrollment in Public and Private Schools of the United States (in millions)		
Year	Public Schools	Private Schools
1980	40.9	5.3
1983	39.5	5.7

Source: U.S. Department of Education

Suppose you meet a new friend who attends school in the United States.

1. What is your estimate of the probability that he or she attends public school?

2. What would have been your estimate in 1980?

3. Estimate the odds in favor of a pupil attending private school in 1983. Do the same for 1980.

Favorite Sports

A Lou Harris poll of April, 1981, obtained the following results for the question "What is your favorite sport?"

Sport	Percent
Football	36
Baseball	21
Basketball	12
Tennis	5
Auto Racing	5
Golf	4
Boxing	4
Others	13

Source: Reprinted by permission of Tribune Media Services.

If you meet a new friend today:

1. What is the probability that the friend's favorite sport is basketball?

2. What is the probability that the friend's favorite sport is *not* football?

3. What are the odds in favor of the friend's favorite sport being baseball?

4. What are the odds against the friend's favorite sport being tennis?

V. COMPOUND EVENTS

Sometimes an event consists of two or more simple events that are considered together as a single event. For example, an outcome from the toss of two coins (two heads, for example) may be thought of as a single event, even though it consists of outcomes on two different coins. Such an event is called a compound event.

Application 23

Tossing Two Coins

    A. Select two coins of different denominations (say, a nickel and a quarter).

    B. Toss the two coins 50 times. Tally the result of each toss in the boxes below.

Two Heads	One Head and One Tail	Two Tails	Total
			50

    1. In what fraction of the tosses did you obtain two heads? Two tails? One head and one tail?

    2. Were these results expected? Why or why not?

    3. Using the results of this experiment, are the three events "two heads," "two tails," and "one head and one tail" equally likely events?

    4. What is your estimate of the probability of obtaining a head and a tail on the toss of two coins?

1. Listing Equally Likely Outcomes

Let us work out the theoretical probability of each of the events in the two-coin experiment. First, it is important to be sure that all the possible outcomes of such an experiment are known and that they are equally likely outcomes. Only then can we use the formula to calculate the theoretical probability of various events.

In the case of the two-coin experiment, either coin may fall heads (H) or tails (T). To keep track of exactly what may happen, it is convenient to work with different kinds of coins—say, a quarter and a nickel. The quarter can come up heads or tails. If the quarter comes up heads, the nickel can come up heads or tails. Similarly, if the quarter comes up tails, the nickel can be either heads or tails. This situation can be shown by means of the accompanying *tree diagram.*

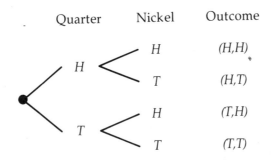

It is convenient to think about such compound events in sequence—as though one thing follows the other. In this case, the first thing (toss of the quarter) can happen in two ways, the second thing (toss of the nickel) can happen in two ways, and the first thing followed by the second thing can happen in 2 × 2 or 4 ways. There are four branches of the tree, and each branch corresponds to exactly one of the four equally likely outcomes. Thus we see that *(H,T)* and *(T,H)* are two distinctly different outcomes. The outcomes would be different even if the two coins were the same kind (both quarters, for example).

The probability of the event *E*, obtaining one head and one tail on the toss of two coins, can now be calculated by using the formula $P(E) = \frac{f}{n}$.

$$P(E) = \frac{2}{4} = \frac{1}{2}$$

In other words, it is expected that roughly 50 percent of the time the toss of two coins will result in one head and one tail. Look back at the two-coin experiment. Did you find that about one half (or 50 percent) of the tosses resulted in one head and one tail? Were you close? It is unlikely that the result was exactly one half. After all, the theoretical probability of one half is what we expect in the long run. How might you continue the two-coin experiment to obtain a fraction closer to one half?

Application 24

Die and Coin

Suppose you roll a die and then toss a coin. What are the possible outcomes? Complete the tree diagram on page 33 to show the possible outcomes.

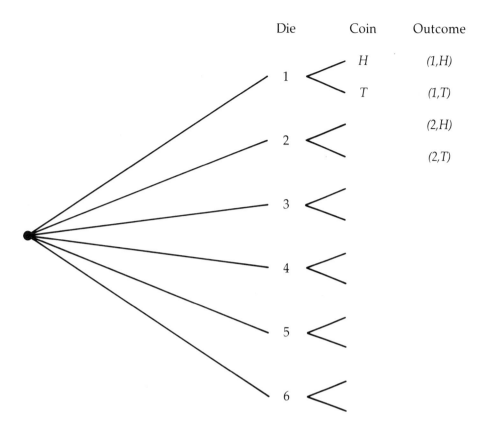

1. How many ways can the die fall?

2. How many ways can the coin fall?

3. How many outcomes are there for the compound experiment "roll a die and then toss a coin"?

4. Do you notice a relationship among your answers to questions 1, 2, and 3?

5. Are the outcomes equally likely?

6. What is the probability of rolling a 4 on the die and tossing a head on the coin?

Consider an experiment that involves tossing two distinguishable dice, say, one red die and one white die. Construct a tree diagram and use it to show that there are 6 × 6 or 36 ways that two dice can fall.

There are other ways of listing the equally likely outcomes in a two-part experiment. For example, a table can be used to display the 36 distinct ordered pairs that represent the 36 possible outcomes of the two-dice experiment.

Red Die

	1	2	3	4	5	6
1	(1,1)	(1,2)	(1,3)	(1,4)	(1,5)	(1,6)
2			(2,3)		(2,5)	
3		(3,2)			(3,5)	
4					(4,5)	
5					(5,5)	
6					(6,5)	

White Die

Notice that (3,2) and (2,3) represent different outcomes. The ordered pair (3,2) represents a 3 on the white die and a 2 on the red die, whereas (2,3) represents a 2 on the white die and a 3 on the red die. Complete the table by recording an ordered pair in each cell of the table.

It is also possible to use an array of points to represent outcomes in a two-part experiment. The accompanying diagram shows a 6 × 6 array of 36 dots (points). The points on the axes are used for reference so that the ordered pair corresponding to each point can be read easily. Points A and B have been named (2,3) and (3,2), respectively. What ordered pairs are represented by points C and D? Locate the point representing the ordered pair (6,6) and label it M.

White Die

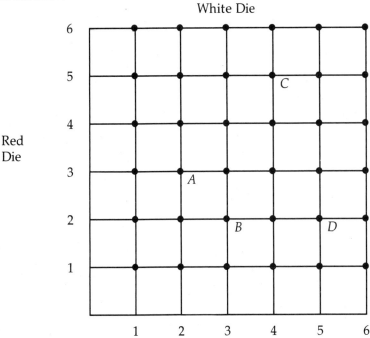

Red Die

Application 25

Two-Dice Experiment

Use the accompanying 6 × 6 array of points to represent the 36 possible outcomes in a two-dice experiment.

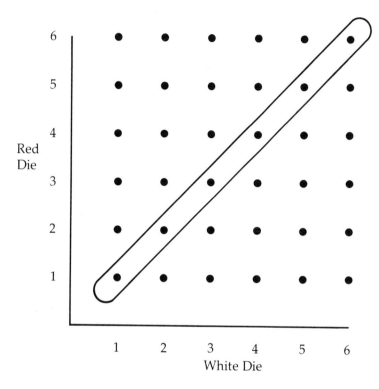

1. A ring has been placed around the points (1,1), (2,2), (3,3), (4,4), (5,5), and (6,6). Calculate the probability of obtaining "doubles" (the same number on each die).

2. Draw a ring around the points that represent the event "the sum of the numbers on the two dice is 8." [Remember that (2,6) and (6,2) are two different favorable outcomes.]

3. Calculate the probability of obtaining a sum of 8 on the toss of two dice.

4. What is the probability of obtaining a sum that is *not* 8?

5. What is the probability of obtaining a sum of 8 and the same number on each die?

6. Suppose that the two dice show the same number. What is the probability that their sum is 8?

2. Multiplication Principle

By using a tree diagram for each of the compound experiments discussed in the previous section, we can list all the equally probable outcomes and determine the total number of outcomes by counting the number of branches of the tree. The results, summarized in the accompanying table, show that the number of branches in a tree diagram can be calculated easily by using multiplication.

Compound Experiment	Number of Outcomes of First Experiment	Number of Outcomes of Second Experiment	Total Number of Outcomes
Toss of two coins	2	2	$2 \times 2 = 4$
Roll of die and toss of coin	6	2	$6 \times 2 = 12$
Toss of two dice	6	6	$6 \times 6 = 36$

Sometimes it is not necessary to construct a complete tree diagram. Instead, multiplication can be used to calculate the total number of outcomes without listing all of them. This Multiplication Principle can be stated as a general rule. If one thing can be done m ways, and a second thing n ways, then there are $m \times n$ ways of doing both things. Thus, for example, if there are three ways of traveling from City A to City B (auto, train, plane) and two ways of traveling from City B to City C (boat, plane), then there are 3×2 or 6 ways of traveling from City A to City C with a stopover in City B.

Application 26

On Your Own

1. Spinner A is divided into six equal sectors, numbered 1, 2, 3, 4, 5, 6. Spinner B is divided into eight equal sectors, numbered 1, 2, 3, 4, 5, 6, 7, 8. In a certain game, spinner A is spun and then spinner B is spun.

   a. How many outcomes are possible?

   b. Are the outcomes equally likely?

   c. What is the probability of spinning a 3 on spinner A and an 8 on spinner B?

2. A certain restaurant offers select-your-own sandwiches. That is, a person may select one item from each of the categories listed:

Bread	Filling	Extras
White Wheat	Tuna Chicken Cheese	Sprouts Lettuce

a. Using a tree diagram, list all possible sandwiches that can be ordered.

b. Would you expect the choices of a sandwich to be equally likely for most customers?

3. A certain General American model car can be ordered with one of three engine sizes, with or without air conditioning, and with automatic or manual transmission.

a. Show, by means of a tree diagram, all the possible ways this model car can be ordered.

b. Suppose you want the car with the smallest engine, air conditioning, and manual transmission. A General American agency tells you there is only one of the cars on hand. What is the probability that it has the features you want, if you assume the outcomes to be equally likely?

4. Jennifer dresses in a skirt and a blouse by choosing one item from each category. Show, by means of a tree diagram, all the outfits she can make.

Skirts	Blouses
Tan Plaid Gray Striped	White Pink Red

3. Multiplying Probabilities: Independent and Dependent Events

Consider the two events *(A)* it rains on Saturday and *(B)* you go to a movie on Saturday. Are *A* and *B* independent? That is, does either event affect the other? If you intend to go to a movie on Saturday, rain or no rain, then *A* and *B* may be independent. If, however, you plan to attend a movie in the afternoon only if you cannot go swimming, then *A* and *B* may be dependent. In this case, your probability of event *B* will change, depending on the outcome of event *A.*

Whenever two coins are tossed, either coin may fall heads or tails, and the way one coin falls does not depend on what happens to the other. The outcome for the nickel is not affected by the outcome for the quarter, and vice versa. In other words, the coin tosses are independent of each other, and the two tosses are said to be *independent events.* Each of the compound events we have been discussing consists of two independent events. The "roll of die" and "toss of coin" are independent events. If we roll two dice, the "roll of the white die" and the "roll of the red die" are independent events.

In a two-coin experiment, the number of equally likely outcomes can be calculated by means of a tree diagram or by using the Multiplication Principle. That is, since each coin has two possible outcomes, the two coins have 2×2 or 4 outcomes. In calculating the probability of each of the four equally likely outcomes, we can use the definition $P(E) = \frac{f}{n}$ and obtain $\frac{1}{4}$. Notice also that the probability of obtaining heads on the nickel is $\frac{1}{2}$, the probability of obtaining heads on the quarter is $\frac{1}{2}$, and the product of these individual probabilities, $\frac{1}{2} \times \frac{1}{2}$, gives us the same result, $\frac{1}{4}$.

Let us see whether this multiplication of probabilities works for other pairs of independent events. Suppose a die is rolled and a coin is tossed, and we are interested in the outcome "roll a 4 on the die and toss heads on the coin." In Application 24, there are 6×2 or 12 equally likely outcomes, and *(4,H)* is one of them. Thus $P(4,H) = \frac{1}{12}$. Notice that $P(4) = \frac{1}{6}$, $P(H) = \frac{1}{2}$, and $\frac{1}{6} \times \frac{1}{2} = \frac{1}{12}$. In other words, $P(4,H) = P(4) \times P(H)$ for these two independent events.

In the case of the toss of two dice, the probability of each of the 36 equally likely outcomes is $\frac{1}{36}$. The probability of obtaining a 2 on the white die is $\frac{1}{6}$, the probability of obtaining a 6 on the red die is $\frac{1}{6}$, and the probability of obtaining (2,6) is $\frac{1}{36}$. Once again, $\frac{1}{6} \times \frac{1}{6} = \frac{1}{36}$, and $P(White\ 2) \times P(Red\ 6) = P(2,6)$. This property is a natural outgrowth of tree diagrams and the Multiplication Principle. For any two independent events, A and B,

$$P(A \text{ and } B) = P(A) \times P(B)$$

In the spinner question in Application 26 (question 1, part c), the probability of obtaining a 3 on Spinner A is $\frac{1}{6}$ (there are six equal sectors), and the probability of obtaining an 8 on Spinner B is $\frac{1}{8}$ (there are eight equal sectors). *Spin A* and *spin B* are two independent events. Therefore, the formula $P(A \text{ and } B) = P(A) \times P(B)$ should give the required probability. That is, $P(3,8) = \frac{1}{6} \times \frac{1}{8}$. There are 6×8 or 48 equally likely outcomes for *spin A* and *spin B*, and (3,8) is one of them. Therefore, $P(3,8) = \frac{1}{48}$ is correct.

Often, however, events are not independent. Suppose a box contains 3 red marbles and 2 blue marbles, all of the same size. A marble is drawn at random. The probability that it is red is $\frac{3}{5}$. If you pick a marble, replace it, and randomly pick again, the probability of picking a red marble remains $\frac{3}{5}$. However, suppose you pick a red marble, do not replace the marble, and then pick another marble. What is the probability of picking a red marble then?

To keep track of the different marbles, let us use r_1, r_2, and r_3 to identify the red marbles, and b_1 and b_2 to identify the blue marbles. Assuming that a red marble has been picked on the first draw, only the following situations could exist:

Select r_1	Leave r_2, r_3, b_1, b_2
Select r_2	Leave r_1, r_3, b_1, b_2
Select r_3	Leave r_1, r_2, b_1, b_2

There would be only two red marbles left, and the bag would now contain only four marbles. The probability of obtaining a red marble on the second draw, assuming that the first pick is a red, is $\frac{2}{4}$ or $\frac{1}{2}$. The second draw is clearly dependent on the first. Check to see that, if the first marble is blue and it is not replaced, the probability of obtaining a red marble on the second draw would then be $\frac{3}{4}$.

Your favorite baseball team is about to play a double header (two games in a row). If your team has won 60 percent of its games in the past, can we simply multiply 0.6 times 0.6 to estimate the probability of the team winning both games? The answer is yes, if we assume the outcomes of the two games to be independent. We often assume events to be independent so that we can approximate probabilities. In the case of your team's double header, the assumption of independent outcomes is probably not far from the truth and gives us an easy way to estimate the probability of the team winning both games.

Application 27

Are They Independent?

In each question below, state whether you think the pairs of events are independent or dependent.

1. *A:* Fred plays with his home video game about one hour a day.

   B: Fred will score more than 2,000 points the next time he plays Pacman.

2. *A:* The Dallas Cowboys will win the Super Bowl this year.

   B: Mount St. Helens will erupt again this year.

3. *A:* Lucy will get an A on her next spelling quiz.

   B: Lucy got an A on her last spelling quiz.

4. *A:* It will snow tonight.

   B: Fred's school bus will be late tomorrow morning.

5. *A:* The next child born in your county will be a boy.

   B: The last child born in your county was a boy.

Marbles in a Box

A box contains 3 red marbles and 2 blue marbles, indistinguishable except for color. A marble is drawn at random, and, without replacing it, another marble is drawn. Complete a tree diagram for this experiment.

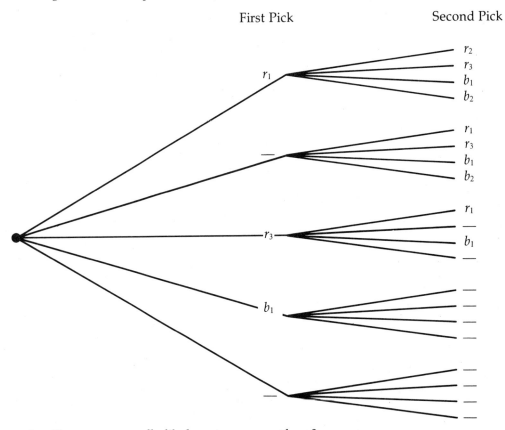

First Pick   Second Pick

1. How many equally likely outcomes are there?

2. Place an asterisk next to those picks that show two successive red marbles. How many of these outcomes are there?

3. a. Compute the probability of drawing two red marbles.

   b. About what percentage of the time can you expect to draw two blue marbles?

   c. Do you have a better chance of drawing a blue marble first and then a red marble or a red marble first and then a blue marble—or are these equally likely events?

   d. What is the probability of drawing a red marble and a blue marble?

Consider the probability of drawing two red marbles from the box of marbles in Application 28. Using the tree diagram or the Multiplication Principle, we see that there are 5×4 or 20 equally likely outcomes. Of these, there are 3×2 or 6 outcomes showing two red marbles. Therefore,

$$P\text{(two reds)} = \frac{3 \times 2}{5 \times 4} = \frac{6}{20} = \frac{3}{10}$$

Here, too, the probabilities can be multiplied if we are careful to observe how the second draw depends on the first draw. We can see that

$$P\text{(red on first draw)} = \frac{3}{5}$$

and

$$P\text{(red on the second draw, given that a red was taken out on the first draw)} = \frac{2}{4} = \frac{1}{2}$$

The probability of observing two reds can be written as:

$$P\text{(two reds)} = P\text{(red on the first draw)} \times P\text{(red on the second draw,}$$
$$\text{given that a red was taken out on the first draw)} = \frac{3}{5} \times \frac{2}{4} = \frac{6}{20} = \frac{3}{10}$$

In the case of the 3 red marbles and 2 blue marbles, verify that

$$P\text{(two blues)} = \frac{2}{5} \times \frac{1}{4} = \frac{2}{20} = \frac{1}{10}$$

$$P\text{(blue first, red second)} = \frac{2}{5} \times \frac{3}{4} = \frac{6}{20} = \frac{3}{10}$$

$$P\text{(red first, blue second)} = \frac{3}{5} \times \frac{2}{4} = \frac{6}{20} = \frac{3}{10}$$

Drawing a red marble and a blue marble consists of two separate compound events: *(blue first, red second)* or *(red first, blue second)*. The tree diagram for Application 28 shows that there are a total of 12 outcomes in which either draw is red and the other draw is blue. Thus, the probability of drawing a red marble and a blue marble is $\frac{12}{20} = \frac{3}{5}$.

Application 29

On Your Own

1. To prevent cable damage as the result of an overload of circuits, an electric company has installed two special switching devices that work automatically and independently to shut off the flow of electricity when the demand for electricity is too great. Experience shows that the first switch has not worked 25 percent of the time and the second switch has not worked 30 percent of the time. (This can be interpreted to mean that the probability that the first switch will fail is $\frac{25}{100}$, and the probability that the second switch will fail is $\frac{30}{100}$.)

a. What is the probability that both switches will fail to work the next time there is an overload?

b. Approximately what percent of the time do both switches fail?

2. A standard deck of 52 playing cards has 4 suits and 13 cards in each suit. The 52 cards of a deck are displayed in the accompanying array of dots.

Spades . . . .

Hearts . . . .

Diamonds . . . .

Clubs . . . .
2   3   4   5   6   7   8   9   10   J   Q   K   A

a. A card is drawn at random. What is the probability that the card is:

(i) a heart?   (ii) an ace?   (iii) the ace of hearts?

b. Suppose the card is replaced and another card is drawn. What is the probability that the card is:

(i) a heart, if the first card was a heart?
(ii) an ace, if the first card was an ace?
(iii) the ace of hearts, if the first card was the ace of hearts?

c. Suppose the card is not replaced and another card is drawn. What is the probability that the second card is:

(i) a heart, if the first card was a heart?
(ii) an ace, if the first card was an ace?
(iii) the ace of hearts, if the first card was the ace of hearts?

d. If two cards are drawn, what is the probability that they are:

(i) two aces?
(ii) any pair?
(iii) the jack of diamonds followed by the king of clubs?

4. Adding Probabilities

Consider the experiment of a single toss of a standard die. There are six equally likely outcomes: 1, 2, 3, 4, 5, 6. Suppose we define certain events as follows:

A   observe a 2
B   observe a 6
C   observe an even number
D   observe a number less than 5

Each event is a set of one or more of the possible outcomes listed above. That is,

A   = {2}
B   = {6}
C   = {2,4,6}
D   = {1,2,3,4}

Events C and D are shown in the accompanying diagram.

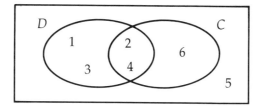

Since the six possible outcomes are equally likely, $P(A) = \frac{1}{6}$ and $P(B) = \frac{1}{6}$. But, what about the probability of observing a 2 or a 6? Two of the six equally likely outcomes are included in the event "observe a 2 or a 6," and so

$$P(observe\ a\ 2\ or\ a\ 6) = P(A\ or\ B) = \frac{2}{6}$$

In this case,

$$
\begin{aligned}
P(A\ or\ B) &= P(A) + P(B) \\
&= \frac{1}{6} + \frac{1}{6} \\
&= \frac{2}{6} \\
&= \frac{1}{3}
\end{aligned}
$$

Will this be true for any two events? Consider the events C and D, as defined above. It is important to recognize that the event $(C\ or\ D)$ includes all the outcomes in C or D or both. That is,

$$
\begin{aligned}
P(C\ or\ D) &= P(observe\ an\ even\ number\ or\ a\ number\ less \\
&\quad\ than\ 5) \\
&= P(observe\ 2,\ 4,\ 6\ or\ observe\ 1,\ 2,\ 3,\ 4)
\end{aligned}
$$

Every outcome except 5 is included in $(C\ or\ D)$. Thus, there are exactly five *different* favorable outcomes, even though the outcomes 2 and 4 are listed twice. Thus,

$$P(C\ or\ D) = \frac{5}{6}$$

But, $P(C) + P(D) = \frac{3}{6} + \frac{4}{6}$, and this sum is $\frac{7}{6}$, not $\frac{5}{6}$. So, $P(C \text{ or } D) \neq P(C) + P(D)$. In fact, $P(C) + P(D)$ cannot represent a probability at all since $\frac{7}{6}$ exceeds 1.

A little investigation will explain what seems to be an inconsistency. Although there are three outcomes in C and four outcomes in D, there are a total of five outcomes in either C or D or both. The outcomes 2 and 4 are contained in both C and D, and care must be taken to count each outcome exactly once. By calculating $P(C) + P(D)$, the probabilities for these two outcomes are added twice. To arrive at the correct result, the probability of this overlap, the event C and D, must be subtracted. That is,

$$P(C \text{ or } D) = P(C) + P(D) - P(C \text{ and } D)$$
$$= \frac{3}{6} + \frac{4}{6} - \frac{2}{6}$$
$$= \frac{5}{6}$$

This result agrees with the probability we originally computed for the event C or D and leads to the Addition Rule for any event of the form C or D:

$$P(C \text{ or } D) = P(C) + P(D) - P(C \text{ and } D)$$

The Addition Rule will work for any two events linked by *or*. Consider the events A and B described above. Does $P(A \text{ or } B) = P(A) + P(B) - P(A \text{ and } B)$? In this case, it is impossible to observe the event A and B (a 2 and a 6 at the same time). Therefore, $P(A \text{ and } B) = 0$, making $P(A \text{ or } B) = P(A) + P(B)$, or $\frac{1}{3}$, as before.

Consider another example of the use of the Addition Rule. Two symptoms of a common disease are a fever *(F)* and a rash *(R)*. A local doctor reports that, out of a typical group of ten people with the disease, two will have the fever alone, three will have the rash alone, and one will have both symptoms. The accompanying diagram shows this situation. (The integers in the diagram represent the number of people in each set.)

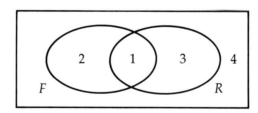

For a randomly selected person with the disease, $P(F) = \frac{3}{10}$, $P(R) = \frac{4}{10}$, $P(F \text{ and } R) = \frac{1}{10}$, and $P(\text{neither symptom}) = \frac{4}{10}$. What is the probability that this person will have at least one symptom—that is, either one or both of the symptoms, the fever *or* the rash *(F or R)*? Now,

$$P(F \text{ or } R) = P(F) + P(R) - P(F \text{ and } R)$$
$$= \frac{3}{10} + \frac{4}{10} - \frac{1}{10}$$
$$= \frac{3}{5}$$

This calculation can be verified by looking at the diagram to see that six out of ten cases fall inside at least one of the sets identifying *F or R*. It is also interesting to notice that "at least one" means "not none." As we have seen, *P(neither symptom)* = $\frac{4}{10}$. Therefore, *P(not neither symptom)* = $1 - \frac{4}{10}$, or $\frac{6}{10}$. The double negative "not neither" is equivalent to "at least one."

In the previous examples, it was possible to find the probabilities without using the Addition Rule. Is the rule really necessary? In some cases it is. One such situation follows.

A computing center has two mainframe computers that operate independently. Let *A* denote the event that computer I is in operation at a randomly selected instant, and let *B* denote a similar event for computer II. Suppose *P(A)* = 0.9 and *P(B)* = 0.8. When you call the computer center from your terminal, what is the probability that at least one computer is running? The event *at least one computer is running* is *A or B*. Using the addition rule for *P(A or B)*, we have

$$P(A \text{ or } B) = P(A) + P(B) - P(A \text{ and } B)$$

Since the computers operate independently,

$$\begin{aligned} P(A \text{ and } B) &= P(A) \times P(B) \\ &= (0.9)(0.8) \\ &= 0.72 \end{aligned}$$

Then,

$$\begin{aligned} P(A \text{ or } B) &= 0.9 + 0.8 - 0.72 \\ &= 0.98 \end{aligned}$$

Note that the probability that at least one computer is in operation is greater than either *P(A)* or *P(B)*. Why is this a reasonable result?

Sometimes the interpretation of numbers in tables requires the correct addition and subtraction of probabilities. The table below shows the race and location of residence of 203,000,000 Americans.

Where Americans Live (in millions)			
	White (W)	Nonwhite (N)	Total
Urban (U)	101	17	118
Suburban (S)	28	3	31
Rural (R)	49	5	54
Total	178	25	203

Source: *The World Almanac & Book of Facts*, 1979 and 1984 edition, copyright © Newspaper Enterprise Association, Inc. 1978 and 1984, New York, NY 10166.

Suppose a pollster chooses a person at random from the U.S. population. What is the probability that he or she is from an urban or suburban area? Now,

$$P(U \text{ or } S) = P(U) + P(S) - P(U \text{ and } S)$$

But, since a person cannot be from both an urban and a suburban area, $P(U \text{ and } S) = 0$. Thus,

$$P(U \text{ or } S) = P(U) + P(S) - 0$$
$$= \frac{118}{203} + \frac{31}{203}$$
$$= \frac{149}{203}$$

What is the probability that the randomly selected person is from an urban or suburban area if he or she is known to be white? Looking at the column for White,

$$P(\text{urban white or suburban white}) =$$
$$P(\text{urban white}) + P(\text{suburban white}) - P(\text{urban and suburban white})$$
$$= \frac{101}{178} + \frac{28}{178} - 0$$
$$= \frac{129}{178}$$

What is the probability that the randomly selected person is either white or from an urban area? Once again the Addition Rule applies. Noting that 101,000,000 are both white *and* live in an urban area,

$$P(W \text{ or } U) = P(W) + P(U) - P(W \text{ and } U)$$
$$= \frac{178}{203} + \frac{118}{203} - \frac{101}{203}$$
$$= \frac{195}{203}$$

Now try some of the following Applications on your own.

Application 30

Using the Addition Rule

1. Refer to the two-dice experiment in Application 25. Calculate each probability in two ways, by counting dots and by using the Addition Rule.

   a. What is the probability of obtaining a sum of 8 or doubles?

   b. What is the probability of obtaining a sum of 7 or a sum of 11?

   c. What is the probability of obtaining at least one 6?

2. Refer to the standard deck of 52 playing cards described in Application 29, question 2. Use the Addition Rule to calculate the following probabilities.

   a. What is the probability that a card drawn at random is a heart or a spade?

   b. What is the probability that a card drawn at random is a heart or an ace?

   c. What is the probability that a card drawn at random is a red king or a diamond?

Application 31

Who Sees the Advertisement?

An advertising firm reports that a certain advertisement on regional radio and television is heard by 30 percent of the regional population on radio and is viewed by 20 percent of the population on TV. Only 5 percent of the regional population hear the ad on both radio and TV. These data are displayed in the figure below.

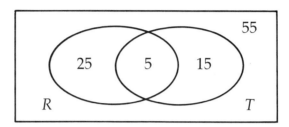

R = Radio, T = Television

Note that 25 percent hear the ad on radio only, and 15 percent see it on TV only.

A customer walks into the store that is sponsoring the advertisement. Find the probability that the customer:

1. has seen the ad on TV.

2. has seen the ad on TV *and* heard it on radio.

3. has neither seen nor heard the ad.

4. has either heard the ad on radio *or* seen it on TV.

What Is the Most Important Subject?

A Gallup youth survey (*Gainesville Sun*, February 13, 1985; The Associated Press) asked a sample of high school juniors and seniors the following question:

What course or subject that you have studied in high school has been the best for preparing you for your future education or career?

Among the males, 30 percent answered "mathematics." Among the females, only 19 percent gave this answer. Answer the following questions under the assumption that approximately 50 percent of the high school juniors and seniors are male. We will let M denote male, F denote female, and A denote the mathematics answer. Thus, MA will denote that a student is male and answered "mathematics."

For a randomly selected high school junior or senior, find the probability that:

1. The student answered "mathematics," given that she is female *(A given F)*.

2. The student is a female and answered "mathematics" *(FA)*.

3. The student answered "mathematics" *(A)*. (Note that $A = MA$ or FA.)

4. The student was male, given that the student answered "mathematics" *(M given A)*. [*Hint:* This is the ratio of the number of males who answered "mathematics" to the total number who answered "mathematics."]

Application 33

How Americans Get to Work

According to a recent survey (*Gainsville Sun*, February 25, 1985), 64 percent of Americans get to work by driving alone. Other methods for getting to work are listed in the table below.

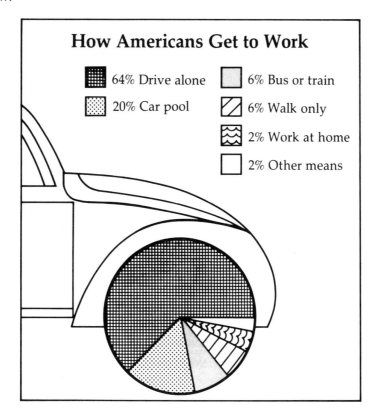

How Americans Get to Work

- 64% Drive alone
- 20% Car pool
- 6% Bus or train
- 6% Walk only
- 2% Work at home
- 2% Other means

1. What is the probability that a randomly selected worker drives alone or in a car pool to get to work?

If two workers, *A* and *B*, are selected independently, what is the probability that:

2. both drive alone?

3. *A* drives alone and *B* walks?

4. *B* drives alone and *A* walks?

5. one of the two drives alone and the other walks?

6. exactly one of the two drives alone?

7. at least one of the two drives alone? [*Hint:* "At least one" means "exactly one of the two *or* both."]

Measuring Association

Foresters and biologists are often interested in studying how two species of plants (or animals) either mingle with each other or separate from each other. Suppose that we are studying two species of trees, A and B, that are both growing in a forest. One method of measuring their association is to randomly sample a tree, observe its species, and then observe the species of its nearest neighbor. This process is repeated for many trees. The data are then recorded as shown below. (The data represent experiments in two different forests.) Use these data to answer the questions below.

Species of Nearest Neighbor

Species of Sampled Tree

	A	B	Total
A	30	10	40
B	5	55	60
Total	35	65	100

Forest I

Species of Nearest Neighbor

	A	B	Total
A	5	35	40
B	30	30	60
Total	35	65	100

Forest II

1. For a randomly selected tree from Forest I, find the probability that:

   a. It is of species A.

   b. It has a neighbor of species A, given that it was observed to be of species A.

   c. Both it and its neighbor are of species A.

   d. Both it and its neighbor are of the same species.

2. Answer question 1, parts a through d, for a randomly selected tree from Forest II.

3. Do you think the probability found in question 1, part d, measures association? Explain.

4. Which forest seems to have more mixing of the species? Which forest seems to have more separation of the species? Discuss how you formulated your answers.

VI. SUPPLEMENTARY APPLICATIONS

The following Applications review the ideas presented in the first five sections of this book.

Application 35

Counting M&Ms

Open a bag (any size) of M&M candies. Make a list of all of the colors represented (dark brown, light brown, yellow, etc.). Next to each color listed, write the number of candies in the bag that are that color (example: yellow, 4; light brown, 7). Be sure to save this list for later.

1. How many M&Ms were there all together?

2. If we put all the candies back into the bag and select one from the bag without looking, what color is it most likely to be? What color is it least likely to be?

3. Is your answer to question 1 an even number or an odd number? If it is an odd number, select one M&M from the bag and eat it!

4. You should now have an even number of M&Ms. Randomly divide them into two piles of equal size. How many are there in each pile? (*Random* means that the M&Ms should be divided into two piles without paying attention to the colors.)

5. Without looking at the piles, which pile is likely to have more yellow candies?

6. Now take turns matching one M&M from pile 1 with one of the same color from pile 2. Remove the two candies of the same color and eat them. There will be some candies in each pile that have no color match in the opposite pile. Look at their colors. Can you explain why these colors are left?

7. Look back to question 5 above. If the division into two piles was carried out randomly, then neither pile is more likely to have more yellow candies. Can you explain why?

If you had reached into the original bag of M&Ms and selected one at random (without looking at the color),

8. Find *P(selecting a yellow).*

9. Find *P(selecting a brown).*

10. Find *P(selecting a blue).*

11. Are the answers to questions 8, 9, and 10 theoretical or estimated probabilities?

The Labor Force

Selected Characteristics of the Civilian Labor Force
and the Unemployed, 1982

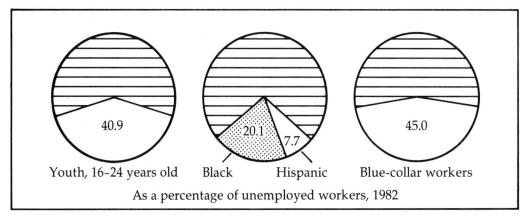

As a percentage of unemployed workers, 1982

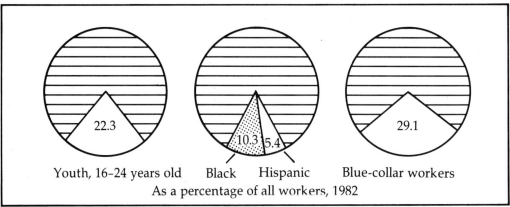

As a percentage of all workers, 1982

Source: "Workers Without Jobs: A Chartbook on Unemployment,"
Bureau of Labor Statistics, 1983.

Use the data in the graphs above to answer the following questions.

1. If you meet an unemployed worker, what is the probability that he or she is black?

2. If you meet a worker, what is the probability that he or she is black?

3. If you meet an unemployed worker, what are the odds in favor of him or her being between 16 and 24 years old?

4. If you meet a worker, what are the odds in favor of him or her having a blue-collar job?

Application 37

Using Random Numbers

Suppose Bob, Mary, Tom, Bill, and Alice are nominated for school safety patrol, but only two are to be chosen. The teacher could randomly choose two from this group of five by:

A.   assigning each student a different number from 1 to 5.

B.   entering the random number table at any place and reading right, left, up, or down until two numbers between 1 and 5 are found.

C.   choosing the two people whose numbers are found in the table.

For example, if we begin in the lower right corner of the random number table in Application 19 and read to the left, the numbers are 0, 5, 9, 3, 2. The first two digits between 1 and 5 are 5 and 3. Thus, students having those two numbers would be selected.

Suppose the five students are numbered from 1 to 5 in the order in which their names appear above. Two students are selected by entering the random number table given in Application 19 at a random point and reading the numbers until the first two distinct digits between 1 and 5 appear. The two students with those numbers are selected.

1.   Draw a tree diagram to represent the outcomes of this selection process. The first column of branches should indicate the five possibilities for the first selection. The second column of branches should indicate the possibilities for the second selection.

2.   How many possible paths are on the tree?

3.   Are the possible outcomes equally likely?

4.   What is the probability that Mary and Tom are selected (in either order)?

5.   What is the probability that two boys are selected?

6.   What is the probability that at least one boy is selected?

7.   What are the odds against selecting two girls?

8.   What are the odds in favor of selecting at least one girl?

Choosing Students

Five students, Art, Bonnie, Carol, Doug, and Ed, volunteer to sell refreshments at the faculty–student basketball game, and only three students are needed. In order to select three students, the following procedure is to be used. Each of the ten possible selections of three students (listed below) is written on a piece of paper, and then one piece of paper is selected at random.

Art, Bonnie, Carol	Art, Bonnie, Doug	Bonnie, Carol, Doug
Art, Bonnie, Ed	Art, Carol, Doug	Bonnie, Carol, Ed
Art, Carol, Ed	Art, Doug, Ed	Bonnie, Doug, Ed
	Carol, Doug, Ed	

1. What is the probability that Art is selected?

2. What is the probability that Doug is *not* selected?

3. What is the probability that both Art and Ed are selected?

4. If Bonnie is selected, what is the probability that Carol is *not* selected?

5. What is the probability that either Bonnie or Carol is selected?

Application 39

Getting a Hit

Mickey Ruth is coming up to bat. The record of outcomes for his last 100 times at bat is shown in the accompanying table. Use these data to answer the following questions.

Home runs	4
Triples	1
Doubles	9
Singles	16
Walks	8
Sacrifices	2
Outs	60
Total	100

1. What is the probability that Mickey will get a hit on his next time at bat?

2. What is his batting average? (The batting average does not count walks or sacrifices as official times at bat.)

3. What is the probability that he will get a home run on his next time at bat?

4. How many walks can he be expected to get out of his next 15 times at bat?

5. What is the probability that he will *not* get a walk on his next time at bat?

6. If he gets a hit, what is the probability that it will be a single?

7. What are the odds against his getting a hit on his next time at bat?

8. What are the odds in favor of his getting an extra-base hit on his next time at bat?

Suppose Mickey Ruth comes up to bat twice in a game.

9. What is the probability that he gets two hits?

10. What are the odds in favor of him getting at least one hit?

Chances of Failing

The diagram below shows the following information. In a certain class of 180 students, all of whom took both English and history, 15 failed history, 10 failed English, and 5 failed both. Find the probability that a student chosen at random from this class:

1. failed history and passed English.

2. failed English and passed history.

3. failed both subjects.

4. failed at least one subject.

5. failed neither subject.

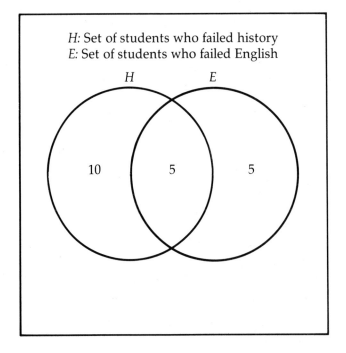

H: Set of students who failed history
E: Set of students who failed English

Application 41

Who Buys the Records?

A Gallup youth survey of 1983 reports that the percentage of teenagers who own over 50 records has decreased from 24 percent in 1981 to 13 percent in 1983. The percentage who own 20 or fewer records increased from 38 percent to 45 percent during the same period. (Source: The Associated Press.)

Suppose there were 1,000 teenagers in your school in 1981.

1. How many would you expect to own over 50 records?

2. How many would you expect to own 20 or fewer records?

3. Answer questions 1 and 2 for a school of 1,000 teenagers in 1983.

4. Does the information from the survey suggest that fewer records were sold in 1983 than in 1981?

5. Suppose you met a teenaged friend in 1983. What are the odds in favor of this friend owning over 50 records?

Probability in the Courtroom

A California court convicted a couple of a crime because the couple had six characteristics observed by witnesses of the crime. The prosecution's case relied heavily on the probabilities of these events, which were approximated as follows:

Characteristic	Probability
Black man with beard	$\frac{1}{10}$
Man with moustache	$\frac{1}{4}$
Woman with blond hair	$\frac{1}{3}$
Woman's hair in ponytail	$\frac{1}{10}$
Interracial couple	$\frac{1}{1,000}$
Driving yellow car	$\frac{1}{10}$

The prosecution argued that the probability of any one couple having all these characteristics was

$$\frac{1}{10} \times \frac{1}{4} \times \frac{1}{3} \times \frac{1}{10} \times \frac{1}{1,000} \times \frac{1}{10} = \frac{1}{12,000,000}$$

Therefore, the couple found to have these characteristics must be guilty.

A higher court overturned this conviction. Would you side with the lower court or higher court? Why? (See *Time*, April 26, 1968, p. 41, for more details.)

Application 43

Straight or Curly Hair?

Two parents with wavy hair can produce a child with hair that is straight, wavy, or curly. Genetic theory states that the odds in favor of such a child having straight hair are 1:3, the odds in favor of wavy hair are 1:1, and the odds in favor of curly hair are 1:3.

Suppose two parents with wavy hair have one child. What is the probability that the child has:

1. straight hair?

2. wavy hair?

3. hair that is not curly?

4. Are the stated odds consistent? That is, can all three statements on odds be true?

Suppose two parents with wavy hair have two children. What is the probability that:

5. both have straight hair?

6. at least one has straight hair?

7. Suppose your school contains 100 students whose parents both have wavy hair. How many of those students would you expect to have curly hair?

Estimating an Area

The square below measures 2 inches on each side, and the circle has a radius of $\frac{1}{2}$ inch. Therefore, the area of the square is 4 square inches, and the area of the circle is $\frac{1}{4}\pi$ square inches.

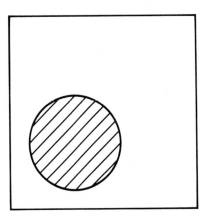

Place the square on a dart board and toss a dart into the square from 8 to 10 feet away. This approximates locating a random point inside the square. Repeat the process until at least 20 darts have landed inside the square. The probability that a dart landing inside the square also lands inside the circle is given by the area of the circle divided by the area of the square.

1. What is the probability of a dart landing inside the circle, given that it has landed inside the square? (Compute this theoretically.)

2. Among the darts landing inside the square, what fraction land inside the circle, for your experiment?

3. Is the experimental proportion in question 2 close to the theoretical probability in question 1?

4. Try the experiment again, and answer questions 2 and 3.

We can use the fraction of darts (random points) falling inside the shaded area to estimate the fractional part of the square covered by that shaded area. In the case of the circle, we knew the area, but for other shapes we might not. Consider the figure on page 61. The area of the square is 4 square inches.

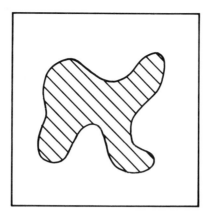

5. Toss darts at the figure from a distance of 8 to 10 feet until at least 20 darts hit inside the square. Of the darts inside the square, how many are inside the shaded area? Calculate the fraction of darts inside the square that are also inside the shaded area. This is an estimate of the fractional part of the square covered by the shaded area.

6. Since the area of the square is 4 square inches, find an estimate of the area of the shaded region.

7. Use this method to estimate an area, or areal proportion, of interest to you. Examples might be the proportion of your state covered by water or the proportion of green in a color photograph.

SPINNER BASE

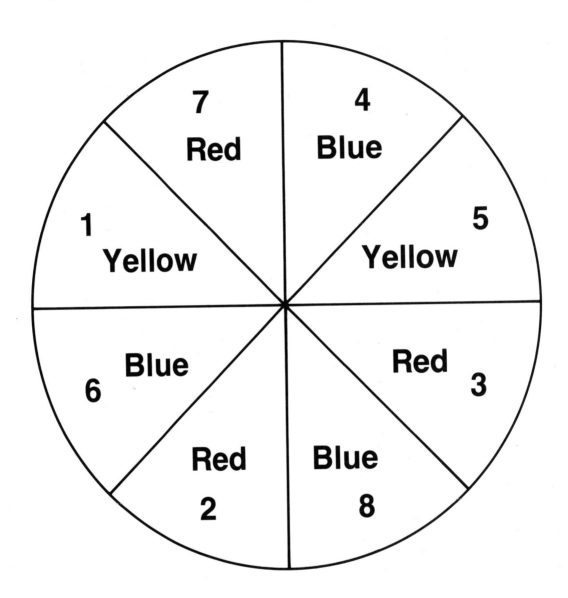

OUTCOMES FOR TOSSING TWO DICE

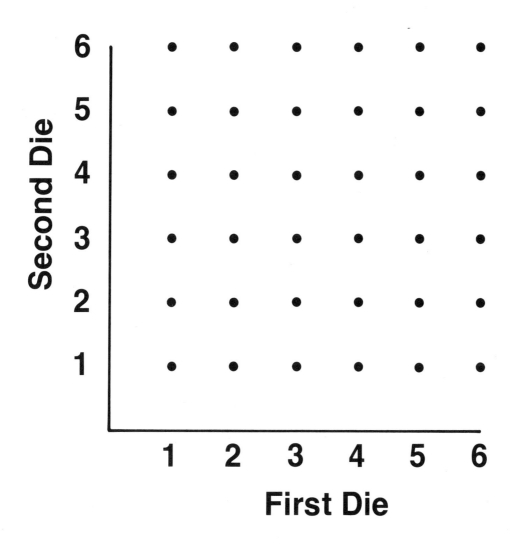